This dinosaur **fossil** is millions of years old.

Fossils

A fossil is the leftover parts of a living thing that died long ago. Evidence left by a long-dead thing can also be a fossil. A footprint can turn into a fossil. These tracks are a kind of fossil. The fossil is called a trace fossil. Over time, the muddy tracks turn into rock.

Dinosaur tracks are trace fossils.

Fossils

Printed in Mexico

ISBN-13: 978-0-15-366648-3

ISBN-10: 0-15-366648-X

2 3 4 5 6 7 8 9 10 805 16 15 14 13 12 11 10 09 08

SCHOOL PUBLISHERS

Visit *The Learning Site!*
www.harcourtschool.com

What Are Fossils?

A **fossil** is what is left of a plant or an animal that lived long ago.

A mold is a kind of fossil. A mold looks like the shape of a living thing carved into a rock. In a mold, there is a space in the rock. The space is the same shape as the living thing.

Sometimes, more mud fills in the space in a mold. Over time, that mud can turn into rock, too. This is called a cast. A cast is a kind of fossil, too. A cast fossil is the opposite of a mold fossil. A cast is the same shape as a plant or an animal. It isn't an empty space.

 What are three kinds of fossils?

Cast

Mold

How Fossils Form

Most fossils are found in sedimentary rock. Sedimentary rock forms when dirt and sand are turned into rock. It takes a long time for sedimentary rocks to form.

When a plant or an animal dies, it might be covered by dirt. Then its soft parts slowly rot and break down. The dirt gets hard and turns into rock. The bones slowly turn into rock. This is how some fossils are made.

 How does a fossil form inside a sedimentary rock?

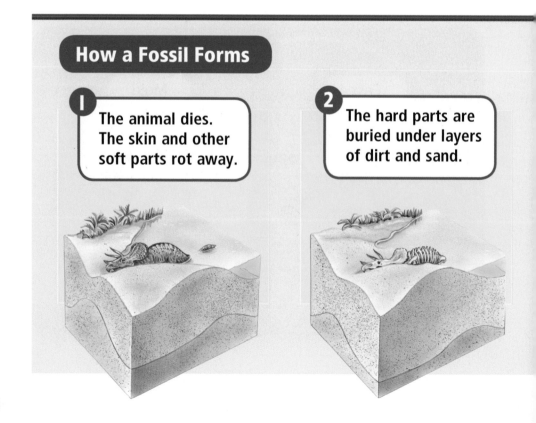

How a Fossil Forms

1 The animal dies. The skin and other soft parts rot away.

2 The hard parts are buried under layers of dirt and sand.

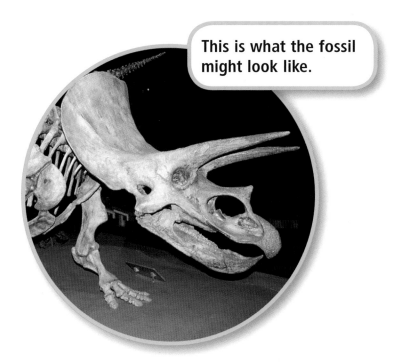

This is what the fossil might look like.

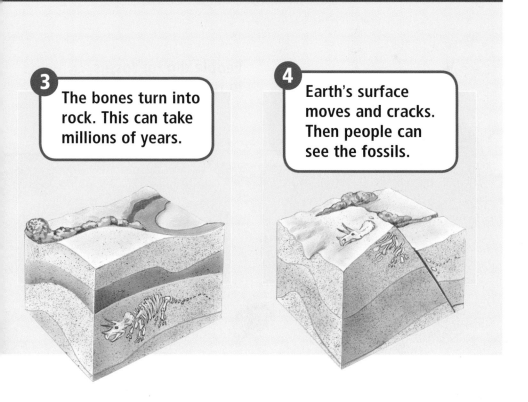

3 The bones turn into rock. This can take millions of years.

4 Earth's surface moves and cracks. Then people can see the fossils.

Learning from Fossils

People study plant and animal fossils to learn about plants and animals that lived long ago. Some plants and animals no longer live on Earth. We know about them because of fossils.

Fossil teeth tell us about animals. Fossil teeth show what an animal ate. Most animals with sharp teeth eat meat. So, if a fossil has sharp teeth, that animal probably ate meat.

People also study rock layers. Each layer was formed at a different time. The deep layers are oldest. Different fossils are in different layers. We can study which plants and animals lived together.

People dig for fossils in rock layers. ▼

People find fossils in Georgia. Very old plant fossils are found in the mountains. Whale and shark fossils are found in southern Georgia. That tells that part of Georgia was under the ocean a long time ago.

 What do sharp teeth fossils tell us?

Fossil shark teeth are sharp.

Review

Complete this main idea sentence.

1. A _____ is the leftover parts of a living thing that lived long ago.

Complete these details statements.

2. Animal tracks are called _____ fossils.

3. A _____ is the space left in a rock when a dead animal breaks down.

4. A _____ forms when mud fills a mold.

What Can We Learn from Fossils?

Dinosaurs are **extinct**.
They no longer live on Earth.

Animals Then and Now

A fossil is the remains of something that lived long ago. Bone fossils have the same shape as animal bones. Teeth fossils have the same shape as animal teeth. Other fossils were left behind by an animal. Footprints are fossils of this kind.

▲ These fossil bones show us how this dinosaur looked.

Some camel fossils are smaller than today's camels. Some camel fossils are larger than today's camels.

Fossils help us learn how animals have changed. We can study fossils of animals that do not live on Earth now. Dinosaurs do not live on Earth anymore. But we can study their fossils.

Some fossils show how animals have not changed. Many camel fossils are like camels that live today. Camel sizes have changed. But camels of the past looked like the camels of today.

 How are past camels like today's camels?

Plants Then and Now

Plant fossils can be harder to find than animal fossils. Plants are softer than animal bones. The soft plant parts rot away too quickly to become fossils.

We can see leaf prints in some rocks. These leaf prints are plant fossils. Ferns are one of the most common plant fossils.

We can also find *petrified* plant fossils. This is when wood turns into a kind of rock.

The fern fossil looks like a fern living today.

Living fern

The ginkgo tree has not changed much in 100 million years.

People can study plant fossils. Fossils help us learn how plants have changed. We can study plants that no longer live on Earth.

Some plants today still look like fossils from long ago. The plants haven't changed over time.

 How are plant fossils like today's plants?

Extinct Plants and Animals

Many kinds of plants and animals are **extinct**. An extinct plant or animal no longer lives on Earth. Dinosaurs are extinct. They no longer live on Earth.

▲ A saber-toothed cat is extinct. This is a saber-toothed cat fossil.

Living things become extinct for different reasons. Sometimes the environment changes. The weather can get too cold or too hot. Some living things change over time when the environment changes. Some animals move to a new place. Other living things cannot change or move. They become extinct.

▲ Fossils show us that saber-toothed cats were shorter than today's lions. But saber-toothed cats were heavier.

North America has had many ice ages. In an ice age, the environment changes. It becomes very cold. Ice covers the land. Many plants and animals become extinct.

After an ice age, the weather gets warmer. Some plants and animals that could live in the cold become extinct when the weather gets warmer.

Woolly mammoths are extinct. They died when the weather became warmer after the last ice age.

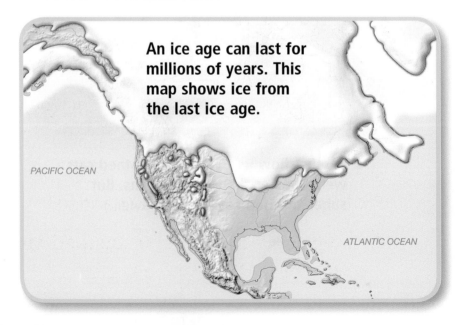

An ice age can last for millions of years. This map shows ice from the last ice age.

PACIFIC OCEAN

ATLANTIC OCEAN

Today, people can cause living things to become extinct. People can change the environment. Plants and animals can lose their homes. They can die and become extinct.

 How was the weather during the last ice age different from today's weather?

The last great auk died in 1844. These birds are now extinct.

Review

 Complete these compare and contrast statements.

1. Some animal _____ are very different from the animals that live today.

2. Some plants became _____ fossils when wood turned into rock.

3. A woolly mammoth is _____ because the ice age environment changed.

4. Fossils show us that saber-toothed cats were shorter than today's _____ .

GLOSSARY

extinct (ek•STINGT) Describes a kind of living
thing that is no longer found on Earth

fossil (FAHS•uhl) The hardened remains of a plant
or an animal that lived long ago